Examples for the
Application of Linear Regression

A CFR Griffin Memorial Publication
No 1.

Examples for the
Application of Linear Regression

Christine Butenuth

Imperial College, University of London, U. K.

and

Gottfried Butenuth

Emeritus Professor

Formerly

Fachhochschule Aachen Juelich, Germany

Tarquin

© The Author s 2005

Published by
Tarquin
99 Hatfield Road
St Albans AL1 4JL

www.tarquinbooks.com

ISBN 1 85853 218 3

All rights reserved. No part of this publication may be reproduced, stored in a retrieval system, or transmitted, in any form or by any means, electronic, mechanical, photocopying, recording or otherwise, without the prior permission of the publisher.

Typeset in the UK

Printed in Great Britain by Lighting Source

CONTENTS

Preface — vii

Section 1

Observations on the Evaluation of Experiments — 1

Section 2

Is a Relationship Univocal? — 5

Section 3

How to Find a Suitable Function — 7

Section 4

Calculation of the "Best Straight Line" by the Least Square Method — 11

Section 5

Use of Approximation Functions — 15

Section 6

Frequently Used Approximation Functions and Their Linearisation — 17

PREFACE

Education in the natural sciences, particularly in practical studies requiring the acquisition of data either inside or outside the laboratory, uses two approaches.

The approach first encountered, normally at school, demonstrates a known fact, e.g. that the extension of an elastic spring has a linear relationship to the load it carries. The purpose of the experiment is to demonstrate a particular aspect of that relationship. This "simple" approach has been very successfully used to teach many scientific laws but it is insufficient for preparing the scientist for practical work at college and later in professional life. Here the relationships observed and measured have to be described in some quantitative form so that they may be studied in ways that allows them to reveal the processes they are reflecting. Thus, the experiment with the spring and a weight could go on to consider what happens to the relationship when the temperature of the spring is changed, when it is loaded beyond the point when it no longer behaves elastically, and when the load applied over a sufficient time for the material from which the spring is made to begin to creep. When attempting to understand such processes it is helpful to do more than just describe the data by a mathematical formula which can reproduce its shape in (x) − (y) space. That provides little basis for explaining the events it describes, thus gives little guidance on what to do to study them further, and provides no basis for extrapolating the data beyond the limits over which it was measured.

At this stage it is far better to search for a quantitative description that fits the data reasonably well and has been used else where in science, engineering or medicine for describing processes that have a known explanation. Such descriptions may not exactly follow all the data, nor exactly follow all the data over its entire plot, i.e. they fit the data approximately, but they have known explanations. These descriptions and their use are called *approximation functions*. Thus one of the main reasons for using an approximation function is to link identifiable processes that may be relevant to the process being studied. For example, a plot of the temperature of a fluid that is cooling with time has definite similarities to a plot of water level with time in a container

The Application of Linear Regression

from which the water is leaking. In both cases something is being lost from storage; heat in the former and volume in the latter. Thus, the selection of a function to use for this purpose is dedicated not only by the shape of the data but also by the processes which are suspected to be occurring: i.e. the selection of the function obliges its user to think about the problem being solved. Such a function is thus a quantitative description of the shape of a plot, where a value for (x) has only one corresponding value for (y), whose form is linked to knowledge of why it may be so. Thus it provides both a means of quantifying data and an insight to the meaning of the data: it puts the data into a larger picture – that of scientific experience. Such a function helps its user understand how to further investigate the process because it suggests possible reasons for the behavior that has been recorded.

Enough is known about many of the processes we observe for us to know the types of relationship to expect and their limits. For example the length of a spring increases as the load upon it increases and the volume of a balloon decreases as the pressure of air inside it decreases towards that of the air surrounding it. These are examples where the experimenter intervenes – adding the load or releasing the pressure: considered over small intervals of time it is possible to observe the spring extending and the balloon shrinking, but these responses may not be of primary interest, the final length and the final volume being the primary points of interest; and here some means of reasonable prediction from the points measured may be required.

Many processes tend towards a time independent value of some sort, as do the spring and balloon described above and often the way in which that time independent value is achieved is of great interest, e.g. a hot body will cool with time to the ambient temperature of its surroundings, and the speed with which that happens will depend on the circumstances – the shape of the body, its insulation and the measurement of the medium surrounding it, to mention just three.

In all these responses it is possible to obtain a function describing the data, using every available data point, but that is usually unnecessary and unproductive. Experience shows that a well selected minimum number of points will provide as useful a function as any: the function is approximate but probably no more "approximate" than a function using more points.

Preface

Further it is advisable to seek and to use functions that contain a small, indeed if possible the smallest, number of terms because some explanation for the constants in the function will eventually have to be provided. For example the relationship between the length of a spring and the load it carries can be described by a proportional where there is one constant which is explained as "elasticity". Closer inspection of the spring vs. load relationship shows that it can also be described by a polynomial, but in that case it is necessary either to have or to want a deeper knowledge of the material from which the spring is made than that which we can describe collectively as "elasticity". If that knowledge does not exist, and is not sought then there may be little advantage in preferring a function containing many terms over one containing a few. Thus polynomial functions containing "n" constants, maxima and minima, inflection points etc. should be used with caution: can they be justified and do the measuring accuracies allow their "existence" to be confirmed?

It is generally good practice to start with a simple function containing as few terms as possible. The price paid for this is that a higher scatter has to be accepted – but this problem can be often treated with the aid of a "best fit method" based on the principles expressed by Gauss.

Some times it is necessary to analyse the functions defined using their differential form, i.e. to quantify the change they describe so as to study whether the change is constant e.g. change in spring length per unit change in the load it carries. Functions requiring this approach may involve a third constant which implies a need for differentiation. The third constant disappears on differentiation, revealing the two remaining constants, whose values permit the third constant to be calculated.

Commonly used steps to establish a reasonable approximation function to data and to quantify its constants will be illustrated by example in the following pages. In every case the example is one where the approximation can be described by either an originally linear function (e.g. the spring length vs. load relationship) or from the constants obtained by linearising an original function e.g. by differentiation (e.g. the temperature of a hot body in cold surroundings, with time) and introducing the original function in it or approximation using a best fit.

The Application of Linear Regression

Such functions are sufficient to describe many of the relationships that are likely to be encountered. It should be remembered that the description of data by a function is rarely an end in it self: usually the function is sought to enable a physical explanation of a response to be provided – that is usually much more difficult a task than finding a function. It also has to be remembered that without a functional description of the experimental data, their physical explanation is not possible.

With regard to the "best fit" lines, it should be remembered that these are based upon Gaussian methods for handling data which use the "least square method". This requires the use of at least six decimal places - a requirement that greatly restricted its use prior to the advent of the calculator: for this reason fit lines were drawn "by eye". There is no need for this nowadays and the best fit should be calculated properly and used routinely: practically every calculator contains instructions showing how this can be done.

As mentioned earlier, finding a function should not be an end in itself: the function is sought because its constants provide a question (what do they represent? what do they mean?) i.e. they help the elucidation of a process. But the process may not be an end in itself because the process helps define a concept. Yet a concept may not be an end in itself – it leads to a decision: "is this so or not?" i.e. it enables questions originally raised by the selection of and by the functions themselves, to be answered.

It is always difficult to formulate a question when you are beyond the area of your expertise, and without a question, no answers can be found! Approximation functions are a reliable aid to asking questions and coming to decisions on an informed basis. It was for that reason this booklet has been written.

SECTION 1

Observations on the Evaluation of Experiments

To experiment is to question: "If this happens will that result?" It is frequently found that in order to answer the initial question many other questions must first be asked and answered. The questions can be qualitative and quantitative: the qualitative question is usually the first to be framed – "Does this happen?": if it does then the quantitative question can be framed so that the amounts of whatever are involved can be measured and compared.

Example: A rock is to be used for construction and aspects of its mechanical properties have to be known. One of these is the relationship of its volume to its temperature.

First the qualitative question is asked: does volume change with temperature? The question compares an *independent variable* i.e., that which induces a change in the other variable or variables (temperature in this case, ϑ in $°C$) with a response that is attached to it, a *dependent variable* i.e., the one in which the change has been induced (volume in this case, V in mm^3). Obviously one is going to be plotted against the other and it is conventional to make the horizontal axis, the abscissa, (x), carry the independent variable and the vertical axis, the ordinate, (y), carry the dependent variable. Once a change in an independent variable has been linked to a change in a dependent variable the nature of that change has to be carefully quantified.

Here the question now becomes strictly *quantitative* - "Does there exist a univocal dependence between the variables?" To be univocal there must be for each (x) value, only one (y) value: thus a univocal relationship would not permit the same volume to exist at two temperatures. This means descriptions such as $y = \pm\sqrt{x}$ are not univocal. A function relates one value of (x) to one value of (y) and only one value of (y): in mathematical notation $y = f(x)$, or in our example $V = f(\vartheta \, °C)$. If a relationship appears to be univocal it is wise to establish whether it is reversible: more will be said about this later – but many natural processes involve a change in the way (x) and (y) are related. Figure 1 illustrates this point. The values in Figure 1a are

The Application of Linear Regression

related in a univocal way where as those in Figure 1b, seen over their whole cycle are not.

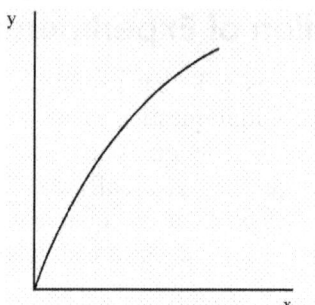

Figure 1a: A function – every x value is unequivocally related to a y value

Figure 1b: demonstrates a relation – x values can be found with two different y values. The relationship is no longer unequivocal.

In nature it is also common to find a number of variables linked to each other so that the relationship of a dependent variable upon an independent one itself depends upon another variable. In the above example volume depends on the temperature, but as the material concerned is to be used in construction it is necessary to know whether the relationship between volume and temperature also depends on the pressure (p in $N \bullet m^{-2}$) at which this relationship is measured. Here pressure and temperature are both independent variables that can change at the same time and quite independently of each other. So, now the function sought contains two variables, i.e. $V = f(\vartheta, p)$: indeed it is often found that independent variables are a function of many independent variables, written $y = f$ (symbols for each variable involved).

When more than one variable is involved the relationship between the dependent and one of the independent variables can only be defined when all other independent variables are held constant at known values. In the above case, the relationship between V and $\vartheta\ °C$ has to be defined for a known and constant (p). Once that has been defined and a function describing it quantified, the pressure (p) may be changed to another

Observations on the Evaluation of Experiments

value and held constant so that $V = f(\vartheta °C)$ can be again determined for that pressure. And so on.

Many experimenters make the mistake of changing more than one variable at a time and leaving too little time in the design of their experimental work for testing each variable systematically: the functions generated from such work are difficult if not impossible to interpret, and thus lose much, if not all, of their worth. Indeed they may be worth less than the paper on which they are written!

SECTION 2

Is a Relationship Univocal?

To prove experimentally that a dependent variable has a univocal relationship with an independent variable it is necessary to proceed as follows: *see Figure 2 below*.

1. Measure (y) with increasing values of (x), points 1, 2, and 3 in Figure 2 and then,
2. Reverse the procedure; points 4, 5 and 6.
3. Once the cycle is completed, sum the values as shown in Figure 2. Mathematically this summation may be represented as

$$\Sigma x_i = (\Delta x \mid_1^2 + \Delta x \mid_2^3 + (\Delta x \mid_3^4 + \Delta x \mid_4^5 + \Delta x \mid_5^6 + \Delta x \mid_6^1)) = 0 \quad (Eq.\ 1.a)$$

4. If the sum is zero (as in this case) the relationship between (y) and (x) is univocal.
5. *Note*: only values for (x) have been used in Eq. 1.a even though values for (y) were also measured. The same approach be used for (y): i.e. $\Sigma \Delta y_i = 0$ (*Eq. 1.b*) and the sum of the Δy_i has to be zero as well. If differentially small steps are considered the integral $\int dy_i$ equals zero as well.
6. If the function behaves like this y(x) is univocal and y is called a *state variable*.

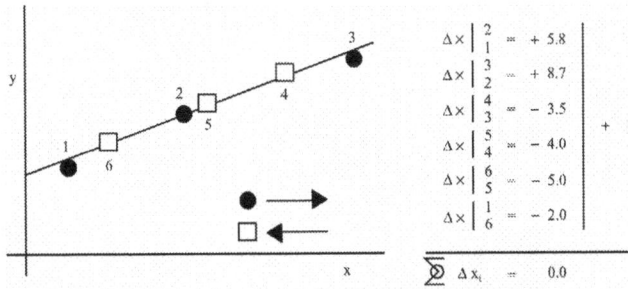

Figure 2: Measurements with increasing x, then the process reversed.

The Application of Linear Regression

Many relationships are found not to be univocal and this is especially true if the dependent variable changes in an irreversible way whilst responding to an independent variable. For example a glass vessel may be heated and then cooled down to its original temperature, and be discovered to have a greater volume than it had to start with! Similar observations have been made in a number of partially crystalline materials e.g. high polymers. Materials which change their density as a result of a process are unlikely to exhibit univocal relationships in the variables involved.

For these reasons it is advisable to measure (y) against (x) using quite large steps at first to obtain an overview of the relationship that is revealed and hence a sense of the function to be sought. It may be possible to use smaller steps on the way back to the origin in order to fill in where the response may be uncertain.

Thus to conclude the example of volume and temperature with pressure: one can write

$$V = V(\vartheta, p)$$

or

$$V = f(\vartheta, p)$$

and when V versus ϑ has been determined under a constant (p) one can write

$$V = V(\vartheta)_p$$

or

$$V = f(\vartheta)_p$$

Always record the conditions under which the relationships were measured.

SECTION 3

How to Find a Suitable Function

EXAMPLE 1: *The determination of two material constants, the specific volume and density of plexiglass (Polymetacrylacidmethylester).*

This example has been chosen to illustrate how a function may be found and to use the simplest of functions: a *proportional*.

Density is mass per unit volume ($\rho \equiv m/V$). Specific volume is its reciprocal, volume per unit mass ($v \equiv V/m$). They are "extensive" magnitudes, i.e. they are magnitudes that depend on the mass of the material. They are thus proportional to the number of unit portions of the material; i.e. they are one or two or x-times greater than their unit mass or unit volume. This means that when mass is plotted vs. volume a straight line is formed which passes through the origin of the coordinate system: i.e. when there is no mass (= no material) there is no volume required for it.

For this reason the mathematical function connecting their values must be a proportional, ($y = m \bullet x$): no other possibility exists and it is worth remembering this even when the measured values may indicate another function. Thought must be given to the limiting values. In the example of the plexiglass, $V = V(m)$ the slope of the function represents the specific volume of the material tested. The points measured are illustrated in Figure 3 overleaf. The scatter of the points can be smoothed out with the aid of the best fit method. The function predicted like this does not pass through the origin (see the large hatched circle). The function found is not a proportional but a straight line i.e.

$$y = m \bullet x \pm c$$

something is "wrong". Here, prior knowledge of the physical nature of the relationship between V and m is invaluable because we know that when $m = 0$, $V = 0$ too. So there must be an error in the measurements; it seems to be systematic as all the points seem to be displaced by the same amount.

The Application of Linear Regression

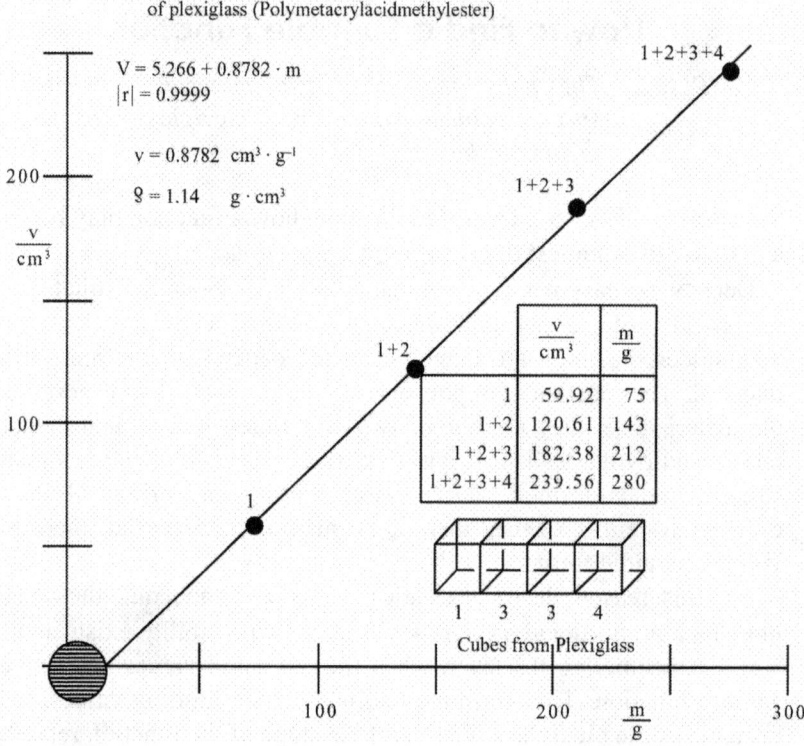

Figure 3: Experimental values of volume versus mass are, in this case, quantitatively described by a linear relationship.

This example shows that it is absolutely essential to identify the physical limits of the relationship being studied before searching for a suitable function to describe it. But it also shows that once a function has been chosen on the basis of physical arguments any deviations from it need to be further explored and explained.

At this point a few proposals for the drawing of diagrams may be helpful:

How to Find a Suitable Function

- For the drawing of overview diagrams it is sufficient to use simple square paper: work with millimeter paper strains the eyes unnecessary. If it is necessary use green millimeter paper for similar reasons: green corresponds to the physiological sensitivity maximum of the human eyes.
- For the sake of clarity the lettering of the co-ordinate axes should be chosen as sparingly as possible: Two numerical values on each axis are sufficient to unambiguously fix their scale. As units of length choose simple multiples: 1 cm, 2 cm or 5 cm etc..
- The size and shape of the symbols which indicate the measuring points should quantitatively reflect the experimental error in both directions. If the error in one direction, for example the abscissa, is essentially smaller than in the other rectangles may be chosen to demonstrate the different range of values in different directions.
- In diagrams used technically it is common to put a net over the picture. This system of co-ordinates facilitates the reading of data later on from unfavorably enlarged or reduced printed diagrams.

SECTION 4

Calculation of the "Best Straight Line" by the Least Square Method

Assume the experimental procedure for measuring a linear connection between dependent and independent variables, has produced the data illustrated in Figure 4.

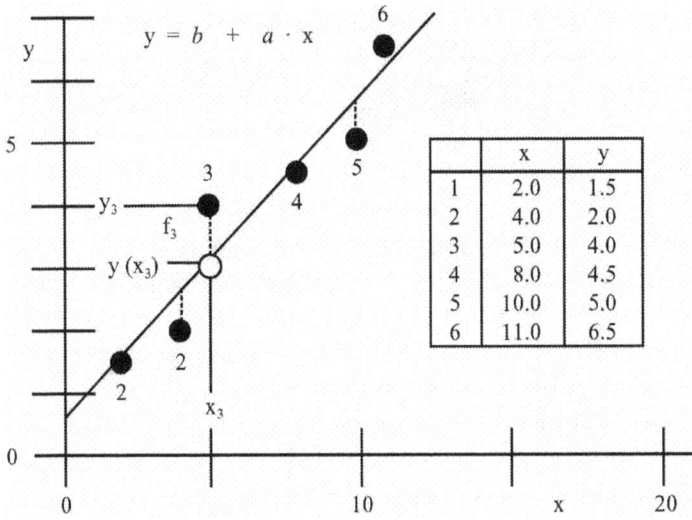

Figure 4: Here the concept of the best fit and what errors "mean" is illustrated.

Six measuring points have been recorded (black circles) P_1 to P_6, of the following co-ordinates x_1/y_1, x_2/y_2.........x_6/y_6. These measured values are fixed numbers and cannot be varied. A straight line to describe their relationship, using a function $y = c + m \bullet x$ is searched for, which will represent the points "as well as possible": this line is also shown on Figure 4. The single measuring points generally do not coincide with this straight line.

The Application of Linear Regression

Their distance from this straight line in the y-direction, for example that of the point 3 on *Figure 4*, be the "error", f_3, of this measured value

$$f_3 = y_3 - y(x_3)$$

or

$$f_3 = y_3 - (c + m \bullet x_3)$$

For mathematical reasons (explained in any good book on statistical methods) the square of the errors, f_i^2 is used.

$$f_i^2 = [y_i - y(x_i)]^2$$

Since all x_i and y_i in Figure 4 are experimentally fixed figures, the error f_i^2 can only change the ordinate intercept, c, and the slope, m, of the straight line: m and c are the variables.

The "best straight line" be that one about which the sum of all error squares, Σf_i^2, is at a minimum. It is located where the deviation of all the points from its slope (m) and the intercept (c) is equal to zero. Thus these derivations are equal to zero and equal to each other.

$$\left(\frac{\partial \Sigma f_i^2}{\partial a}\right)_c = 0 \qquad Eq.\ 2a$$

$$\left(\frac{\partial \Sigma f_i^2}{\partial b}\right)_m = 0 \qquad Eq.\ 2b$$

The algebraic execution of this idea yields the following expressions for the ordinate intercept and slope of the "best straight line".

$$c = \frac{\Sigma y_i \bullet \Sigma x_i^2 - \Sigma x_i \bullet \Sigma(x_i \bullet y_i)}{n \bullet \Sigma x_i^2 - (\Sigma x_i)^2} \qquad Eq.\ 3a$$

$$m = \frac{n \bullet \Sigma(x_i \bullet y_{i+}) - \Sigma x_i \bullet \Sigma y_i}{n \bullet \Sigma x_i^2 - (\Sigma x_i)^2} \qquad Eq.\ 3b$$

Calculation of the "BSL" by the Least Square Method

y_i	x_i	x_i^2	$x_i \cdot y_i$
1.5	2.0	4.0	3.0
2.0	4.0	16.0	8.0
5.0	5.0	25.0	20.0
4.5	8.0	64.0	36.0
5.0	10.0	100.0	50.0
6.5	11.0	121.0	71.5
$\Sigma y_i = 23.5$	$\Sigma x_i = 40.0$ $(\Sigma x_i)^2 = 1600$	$\Sigma x_i^2 = 330.0$	$\Sigma (x_i \cdot y_i) = 188.5$
n = 6			

Table 1: Calculations needed for the best fit curve.

Here n indicates the number of measurements; in the example illustrated (*Figure 4*), n = 6. Table 1 demonstrates the calculations needed for the sums in the above given formulae.

Introducing the single sums into equations 3a and 3b yields m and c of the "best fit line" adopted for the measurements listed:

c = 0.566 and m = 0.503.

This calculated straight line is that shown in *Figure 4*.

If you have not done this before it will be useful to calculate, with the help of your own calculator, the same example and examine the results. Calculate also the sum over all the errors in y-direction.

SECTION 5

Use of Approximation Functions

As demonstrated above, the approximation of measured data by functions presupposes a steady behaviour for the relationship in question. This applies even between single points and means that the function found enables data for each and every place on the graph to be predicted between the limits over which it was measured. Thus approximation functions enable interpolation between measured points.

Such functions enable the first and subsequent derivations at each place to be predicted also, and may be mathematically manipulated to possibly quantify other properties as well which might follow from differentiation.

Functions quantitatively describe graphs and thus allow comparisons to be made between different experiments and the effects or different arrangements in the same experiment. However they do not explain the data. None the less such descriptions can be a great assistance to an interpretation of the data and consideration of the processes it may be reflecting.

SECTION 6

Frequently Used Approximation Functions and Their Linearisation

A number of examples are now given illustrating commonly used functions. Each example uses real data. The value of the examples is not in the specific problem that they solve but in the way they enable a solution to the problem to be obtained. The methods so illustrated can be applied to different problems producing data of similar form. Because the examples are real examples a few words of introduction are often given to enable the reader to understand the nature of the problem being solved.

EXAMPLE 2: Isolation of components seen only as mixtures

Mixtures can be thought of as being of two types: mechanical and ideal (often "chemical"). A *mechanical mixture* is a mixture that can be separated by physical means, e.g. quartz particles mixed up with iron filings. The iron filings can be separated from the quartz grains by using a magnet, for example, i.e. by a physical means. A *true* mixture, or "mixture" is one in which its components are molecularly fine dispersed. Here the components cannot be separated by physical means.

Here is an example of a frequently used linearisation of a function that enables aspects of the individual components to be quantified. The mixture of Fayalite and Forsterite (types of the mineral olivine): this often occurs as an individual mixture, any olivine containing an amount of both.

Let the respective volumes of components 1 and 2 be V_1 and V_2, and the total volume of the mix be V_{total}, then:

$$V_{total} = V_1 + V_2$$

And because $V = m/\rho$ and $v = 1/\rho$ (see *Example 1* in Section 1 where m denotes mass and ρ density):

$V_{total} = v_1 \bullet m_1 + v_2 \bullet m_2$ dividing by $(m_1 + m_2)$, i.e. the total mass

The Application of Linear Regression

$$v_1 \bullet \frac{m_1}{m_1 + m_2} + v_2 \bullet \frac{m_1}{m_1 + m_2}$$

called the average specific volume *Eq 4a*

Let $\quad \dfrac{m_1}{m_1 + m_2} = \xi_i$, called the mass fraction

then $\quad \bar{v} = v_1 \bullet \xi_1 + v_2 \bullet \xi_2 \qquad\qquad Eq.\ 4b$

and because $\xi_1 + \xi_2 = 1$

$\qquad \bar{v} = v_1 + (v_2 - v_1) \bullet \xi_2 \qquad\qquad Eq.\ 4c$

Equation 4c is that for a straight line (y = c + m • x) and so the function can be used in two simple ways:

- To derive the percentage of constituents in a binary mix from knowledge of the densities of each constituent, i.e. a quantitative phase analysis. In *Figure 5* the large half filled circle represents an olivine whose quantitative composition of Forsterite and Fayalite can be read directly from a graph from one measurement, not of its chemical composition, but of its mean specific volume. ($\bar{v} = 0.25$ cm^3 • g^{-1} in this example)
- to derive the numerical value of the densities of the two constituents, pure phases, of a binary mixture from measurements of the specific volume of mixtures containing different proportions of the two constituents. Here the function (*Eq. 4c*) enables the straight line obtained by joining the measured points, to be extrapolated. In *Figure 5* the small open circles illustrate such a case, from which the specific volumes of the two end members can be obtained, $v_1 = 0.2278$ cm^3 • g^{-1} and $v_2 = 0.3049$ cm^3 • g^{-1}.

Example 2 produced data which plots as a straight line but many experiments produce data which plots as a curve: how can these data be treated? Here it is worth considering quadratic functions.

SECTION 6

Frequently Used Approximation Functions and Their Linearisation

A number of examples are now given illustrating commonly used functions. Each example uses real data. The value of the examples is not in the specific problem that they solve but in the way they enable a solution to the problem to be obtained. The methods so illustrated can be applied to different problems producing data of similar form. Because the examples are real examples a few words of introduction are often given to enable the reader to understand the nature of the problem being solved.

EXAMPLE 2: Isolation of components seen only as mixtures

Mixtures can be thought of as being of two types: mechanical and ideal (often "chemical"). A *mechanical mixture* is a mixture that can be separated by physical means, e.g. quartz particles mixed up with iron filings. The iron filings can be separated from the quartz grains by using a magnet, for example, i.e. by a physical means. A *true* mixture, or "mixture" is one in which its components are molecularly fine dispersed. Here the components cannot be separated by physical means.

Here is an example of a frequently used linearisation of a function that enables aspects of the individual components to be quantified. The mixture of Fayalite and Forsterite (types of the mineral olivine): this often occurs as an individual mixture, any olivine containing an amount of both.

Let the respective volumes of components 1 and 2 be V_1 and V_2, and the total volume of the mix be V_{total}, then:

$$V_{total} = V_1 + V_2$$

And because $V = m/\rho$ and $v = 1/\rho$ (see *Example 1* in Section 1 where m denotes mass and ρ density):

$V_{total} = v_1 \bullet m_1 + v_2 \bullet m_2$ dividing by $(m_1 + m_2)$, i.e. the total mass

The Application of Linear Regression

$$v_1 \bullet \frac{m_1}{m_1 + m_2} + v_2 \bullet \frac{m_1}{m_1 + m_2}$$

called the average specific volume   Eq 4a

Let $\dfrac{m_1}{m_1 + m_2} = \xi_i$, called the mass fraction

then $\overline{v} = v_1 \bullet \xi_1 + v_2 \bullet \xi_2$ Eq. 4b

and because $\xi_1 + \xi_2 = 1$

$$\overline{v} = v_1 + (v_2 - v_1) \bullet \xi_2 \qquad Eq.\ 4c$$

Equation 4c is that for a straight line (y = c + m • x) and so the function can be used in two simple ways:

- To derive the percentage of constituents in a binary mix from knowledge of the densities of each constituent, i.e. a quantitative phase analysis. In *Figure 5* the large half filled circle represents an olivine whose quantitative composition of Forsterite and Fayalite can be read directly from a graph from one measurement, not of its chemical composition, but of its mean specific volume. ($\overline{v} = 0.25$ cm^3 • g^{-1} in this example)
- to derive the numerical value of the densities of the two constituents, pure phases, of a binary mixture from measurements of the specific volume of mixtures containing different proportions of the two constituents. Here the function (*Eq. 4c*) enables the straight line obtained by joining the measured points, to be extrapolated. In *Figure 5* the small open circles illustrate such a case, from which the specific volumes of the two end members can be obtained, $v_1 = 0.2278$ cm^3 • g^{-1} and $v_2 = 0.3049$ cm^3 • g^{-1}.

Example 2 produced data which plots as a straight line but many experiments produce data which plots as a curve: how can these data be treated? Here it is worth considering quadratic functions.

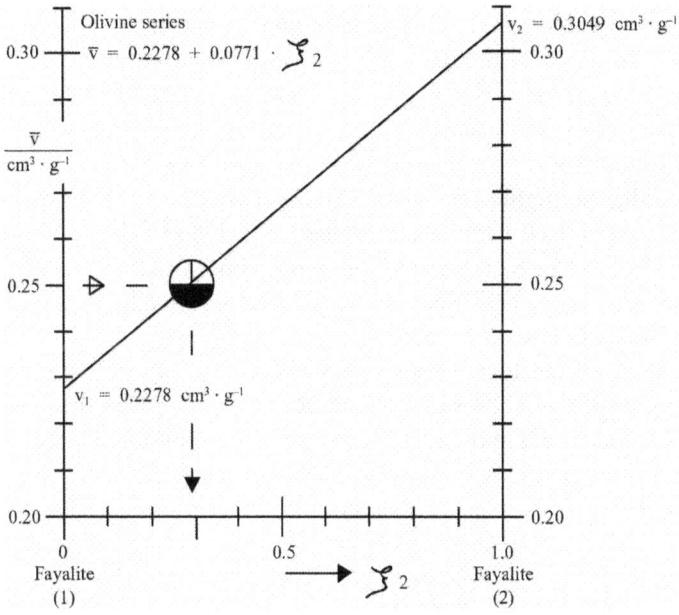

Figure 5: Specific volumes are plotted v the mass fraction of Forsterite.

6.1 Quadratic Functions: Some basic aspects

If the experimentally determined measuring values describe a curve with a significantly marked extreme value – but without inflection points – a quadratic function may be tried as an approximation. Two types of functions are frequently used:

$$y = a + b \bullet x + c \bullet x^2 \qquad Eq.\ 5$$

and

$$(y - y_{\text{extreme value.}}) = a \bullet (x - x_{\text{extreme value}})^2 \qquad Eq.\ 6$$

The Application of Linear Regression

Both forms contain three constants, a, b and c. Such curves can be considered with reference to *Figures 6a* and *6b* and the associated table of values. To linearise the data, and thus an evaluation of the data with the help of the Gauss-method, at least one of these three constants has to be estimated. In *Figure 6a* measured values are given: the constant (a) in *Eq. 5* obviously has its numerical value close to zero; (see the heavy curve and the highlighted point 0/0). In this case $y/x = b + c \cdot x$ may be used as the linearised form and examined with help of the experimental figures. This has been done in *Figure 6b* from which $b = -2.18$ (the intercept) and $c = 1.07$ (the slope of the line). Check *Figure 6b* regarding the x position of the minimum y value.

The second of the formulations above (*Eq. 6b*) may be preferentially used if the position of the extreme can be estimated quite correctly. The two dotted curves are shown in *Figure 6a*, shifted by $a = \pm 2$ compared with the solid curve. Each curve influences each and every y-value of the curve, but $x_{extr.}$ remains unchanged.

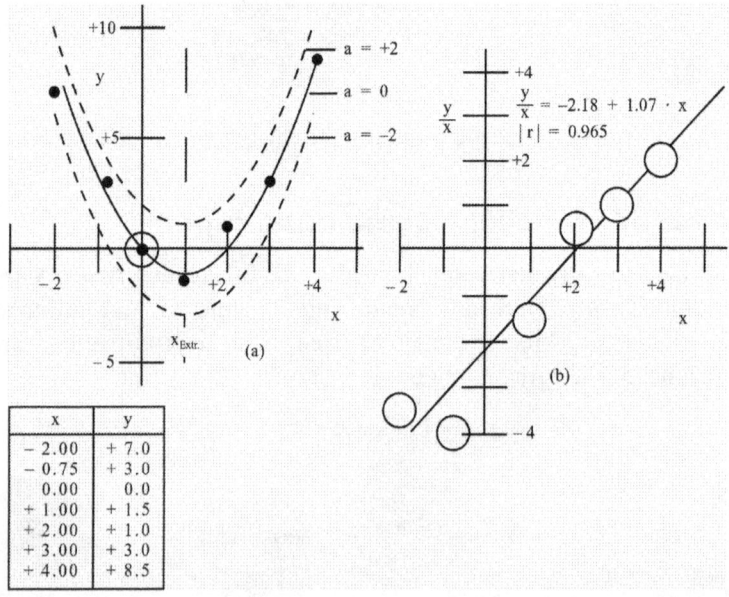

Example 3: Thermal expansion of a liquid: water

Quadratic Functions

How does a body (either solid or liquid) change its dimensions when heated? The approximate answer is as follows:

$$V = V_o + \Delta V (\Delta \vartheta) \qquad Eq.\ 7$$

Where:

V: volume at a given temperature
V_o: volume at the beginning of the experiment
ΔV: difference in volume due to a change in temperature
$\Delta \vartheta$: difference in temperature.

Referring to what was said above in Section 1, the statement made in *Eq. 7* is read as "the volume at any temperature ϑ equals the initial volume, V_o, plus the change in volume that occurs with a change in temperature, $\Delta V (\Delta \vartheta)$, upto the temperature in question." Thus $V = f(\vartheta\ °C)$ in this case. *Eq. 6* may not be a completely adequate statement, as explained in Section 1, but that is what it says and its use presumes that temperature alone affects the change in volume and nothing else.

Think about equations before accepting them.

The function *Eq. 7* is here studied for water, in *Example 4* it is used to study a natural rock salt solution.

The coefficient of thermal volume expansion, is often defined as

$$\alpha \equiv \frac{1}{V_o} \bullet \left(\frac{\partial V}{\partial T}\right)_p \qquad Eq.\ 8$$

T: absolute temperature measured in K, ϑ: temperature measured in °C; in thermodynamics it is usually the absolute temperature that is used.

Eq. 8 says "the coefficient of thermal expansion is equal to the change in volume per unit change in temperature at a given pressure, normalized (i.e. divided by – for reasons explained in Section 3), the initial volume."

The Application of Linear Regression

How may the numerical values for α be obtained?

Here temperature change, $\Delta\vartheta$, and volume change, ΔV, are connected. Such volume changes can be calculated from the variations of the dimensions of the measuring liquid, in the experiment described here the dimension used was the change in height Δh of water, in an ascending pipe, relative to a datum, *Figure 7*. In this experiment the initial temperature was $\vartheta_o = 15°C$, the corresponding volume $V_o = 108.5$ cm^3; radius of the ascending pipe, $r = 0.2$ cm. With these data the calculation of the volume at any temperature is:

$$V = 108.5 + \Pi \bullet (0.2)^2 \bullet \Delta h (\Delta\vartheta) \qquad Eq.\ 9$$

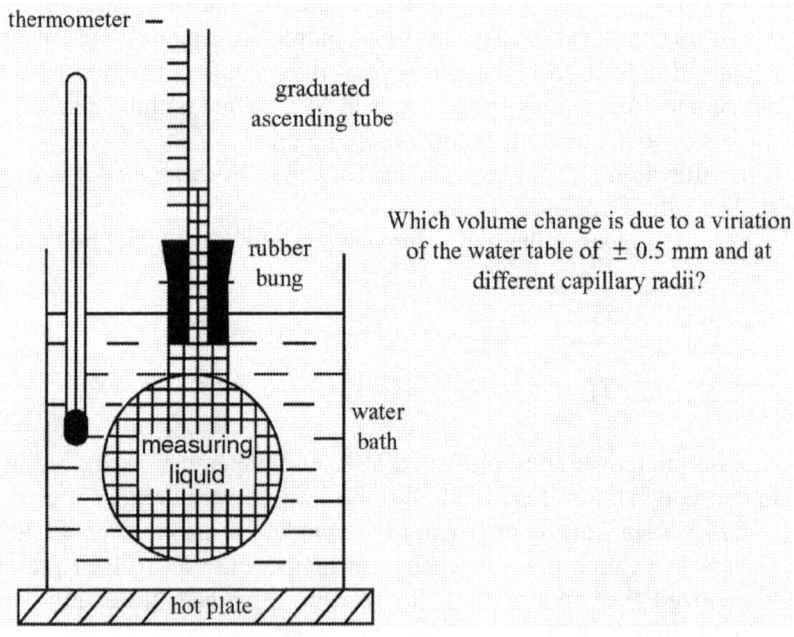

Which volume change is due to a viriation of the water table of $\pm\ 0.5$ mm and at different capillary radii?

Figure 7: The experimental set-up

Quadratic Functions

Experimental measuring values for distilled water are given in *Table 2* and plotted in *Figure 8*.

ϑ/°C	(ϑ – 15)	Δh/cm	V/cm³
15	0	0.00	108.500
20	5	0.35	108.544
25	10	1.25	108.657
30	15	2.35	108.795
35	20	3.70	108.965
40	25	5.25	109.160
45	30	7.15	109.398

Table 2: Experimental values that are needed for the linearization

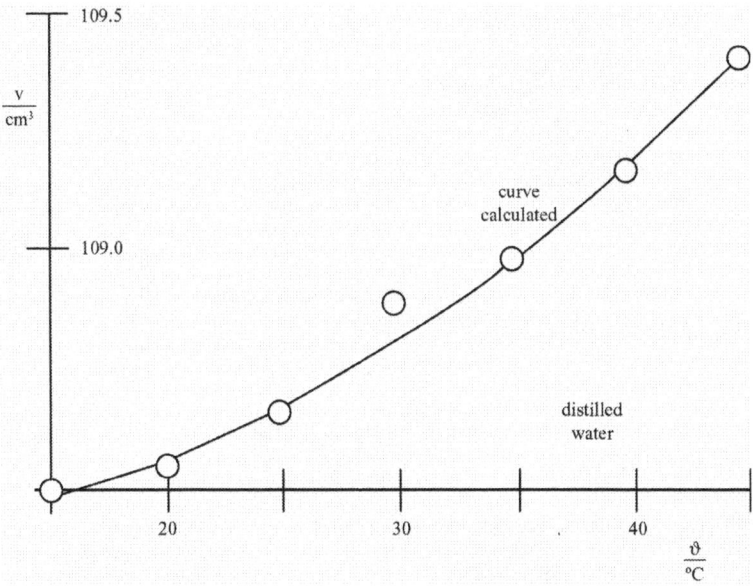

Figure 8: Experimental values of volume plotted against temperature for water.

The Application of Linear Regression

$$V_0 = 108.5 \text{ cm}^3, \vartheta = 15°C$$

The shape of the graph indicates that the measured values should not be described by a linear function (although if you do not think about this, and simply ask the calculator to do it for you, it will!). The nonlinearity may be shown in another way: draw with a pencil a straight line through the first three of the measuring points. It can be easily seen that the following points deviate from this initial line systematically in one direction.

Furthermore, it is known that water shows a minimum of the volume at $\vartheta = 4°C$. From this fact it follows that a function for describing these data must be of the second order (i.e. something ^{2}), a parabola must be suitable. Note how other scientific knowledge is being used here to define the sort of function required (see the Preface). If numbers alone had been considered many students would have attempted to describe – and hence attempted to understand and explain the data – using a straight line!

$$V = V_0 + a \bullet (\vartheta - 15) + b \bullet (\vartheta - 15)^2$$

Thus, the linearised form can be used:

$$\frac{(V - V_0)}{(\vartheta - 15)} = a + b \bullet (\vartheta - 15) \qquad Eq.\ 10$$

In the case of water in this experiment, if the value at 20°C is excluded (always draw a sketch of your results!) equation 10 equals:

$$|r| = 0.999$$

or, after rearranging the previous equation and introducing the numerical value for V_0

$$V = 108.5 + 8.96 \bullet 10^{-3} \bullet (\vartheta - 15) + 7.02 \bullet 10^{-4} \bullet (\vartheta - 15)^2 \Big|_{15\,°C}^{45\,°C}$$

Quadratic Functions

Note that the limits between which this relationship has been measured are cited (15°C and 45°C) because *it is always dangerous to undertake extrapolations beyond the range over which measurements are made!*

At the start of this example a possible definition of the coefficient of thermal expansion was given. It was for a constant pressure, i.e. was isobaric, and thus called the isobaric temperature coefficient of the volume, $(\partial V/\partial T)_p$. It is derived from the function $V(\vartheta)$ (see *Figure 8*):

$$\left(\frac{\partial V}{\partial T}\right)_p = 8.96 \cdot 10^{-3} + 1.404 \cdot 10^{-3} \cdot (\vartheta - 15)$$

(*Note*: the pressure under which this experiment was done was atmospheric pressure, see *Figure 7* for the experimental arrangement.)

A second way of defining the expansion coefficient is not to introduce the fixed reference volume V_o but to use the temperature function $V(\vartheta)$ itself, see Eq. 7. Using this produces a slightly different value:

$$\alpha = 8.26 \cdot 10^{-3} + 1.294 \cdot 10^{-5} \cdot (\vartheta - 15)$$

EXAMPLE 4: The thermal expansion for a saturated rock salt solution

In this experiment a crystalline solid, rock salt (NaCl) is dissolved in water, and the expansion of this aqueous solution measured as a function of temperature. The experimental arrangement is the same as shown in *Figure 7*.

Figure 9 shows that the thermal volume dependence of a saturated rock salt solution. $V(\vartheta)$ is no longer parabolic, but linear. Thus even the minimum at 4°C has disappeared. Therefore the expansion coefficient as defined in *Example 3* is now independent of temperature and therefore constant.

The Application of Linear Regression

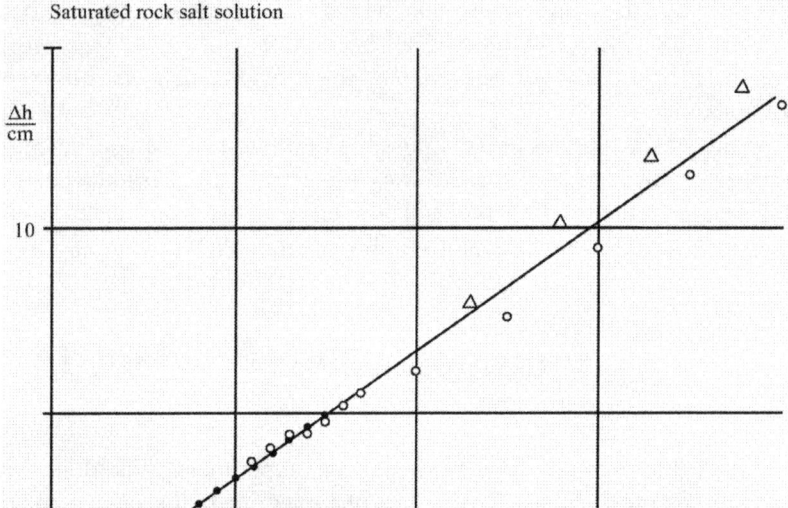

Figure 9: The height of a stand pipe of a saturated rock salt solution is plotted versus its temperature.

The reason for this surprising difference is based on the following molecular facts. The Na^+ and Cl^- ions fix water molecules in the form of so called hydration envelops. In parallel with this the polymer framework structure of the liquid water is destroyed.

The expansion coefficient is a sensitive magnitude of the material.

Quadratic Functions

6.2 More complex relationships

Natural processes may involve complex responses where the identification of a suitable function is not straightforward. Mixtures can create such situations. *Example 2* illustrated an ideal mixture whose specific volume changed in a monotonic way with the composition of the mix (*Figure 5*). Many mixtures are not "ideal"; they are "real mixtures" and do not behave like ideal mixtures. *Example 4* involves such a mix: the function describing the result is not that for water (*Figure 8*), so the presence of NaCl is having an effect – what is happening? Here the unsuitability of the function for water suggests that some other process (or processes) may be operating.

By way of brief and simple explanation, the reason why the function for water does not apply is largely due to the fact that the structure of water is drastically changed by the presence of ionic salts dissolved in it. So two things are changing during the experiment whose results are shown in *Figure 9*: the structure of the solvent *and* the concentration of the solute. These effects can be seen if the volumes involved are considered: this is shown in *Example 5*.

Example 5: Mole volume as a function of mole fraction in a real mixture

A *mole* is a quantity used in chemistry. For the purpose of this example, a molecule (e.g. NaCl, or H_2O) can be considered as a "particle", and a mole is $6.023 \bullet 10^{23}$ particles. Basic text books on chemistry explain why this is so. The moles of a substance therefore possess a volume and a mass, thus a weight and a density. Chemists talk of the mole volumes, mole fractions (i.e. the fraction of a mixture made up by a particular mole type, e.g. the amount of NaCl in a solution of NaCl with H_2O) and mole weights. One mole of Na equals $6.023 \bullet 10^{23}$ particles, the mass of these particles is their mole mass. The formula NaCl describes that one mole of Na combined with one mole Cl creates one mole NaCl.

The Application of Linear Regression

In this example the mean mole volume of the NaCl and H$_2$O mixture ($\overline{V}_0$) is plotted against the mole fraction of the salt in solution (x_{NaCl}): see *Figure 10*, where the values for both are also tabulated.

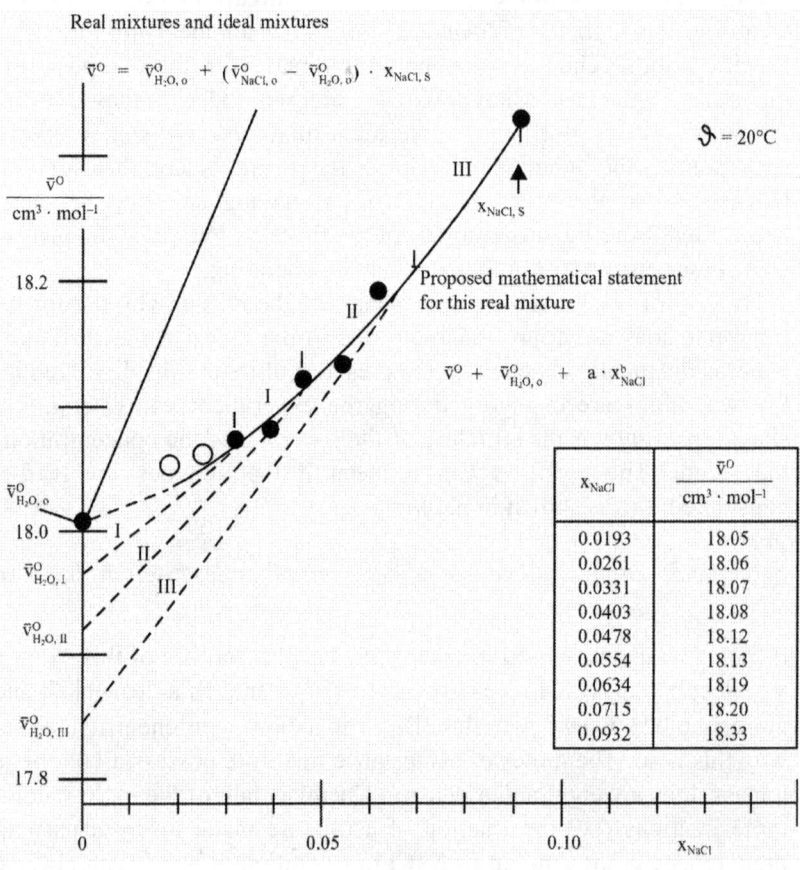

Quadratic Functions

Given these data the first observation is that their relationship is not linear – hence, by reference to the fact that the function for an ideal mixture *is* linear, they cannot be reflecting the behaviour of an ideal solution. Hence again, the use of appropriate functions has enabled *wider knowledge* to be employed.

When searching for a function to describe such data it is necessary to consider whether any physical limits are known about the process being studied (see Section 3). In this case there are two:

- when $x_{NaCl} = 0$, the mean mole volume must be equal to the mean volume of water containing no NaCl. For distilled water this is
$$[]V^o_{H_2O} = 18.01 cm^3 \bullet mole^{-1}$$
- when the solution is saturated, and thus can dissolve no more NaCl, $x_{NaCl} = 0.0932$.

Whatever the function chosen it has to satisfy these criteria. One such function is shown in *Figure 10*.

Problems of this type are amenable by the Lewis and Rendell method (see textbooks on physical chemistry for details). In this case the mole volumes of water and rock salt in solution (called partial mole volumes) can be numerically calculated provided the mathematical function $v_o(x_{NaCl})$ i.e. a function for the graph of the type shown in *Figure 10*, has been obtained.

The approach divides the graph into zones, I, II, III and so on (see *Figure 10*) from which secant lines can be drawn and extrapolated to $x_{NaCl} = 0$. From this it can be seen that in this example the mole volume v^o decreases as the concentration of salt dissolved increases – and the progressive de-structuring of the water quantified in these terms.

Such an approach can be applied to a range of similar problems. Further information on the evaluation of real mixtures or solutions can be found in the textbooks describing the thermodynamics of mixed phases.

The Application of Linear Regression

Example 6: The Langmuir adsorption isotherm: a proportional divided by a straight line

The adsorption of an alcohol, dimethylolcyclohexen, dissolved in decalin on a high dispersed silica, TK-800 of the industry is shown in *Figure 11*. Values are given in *Table 3*.

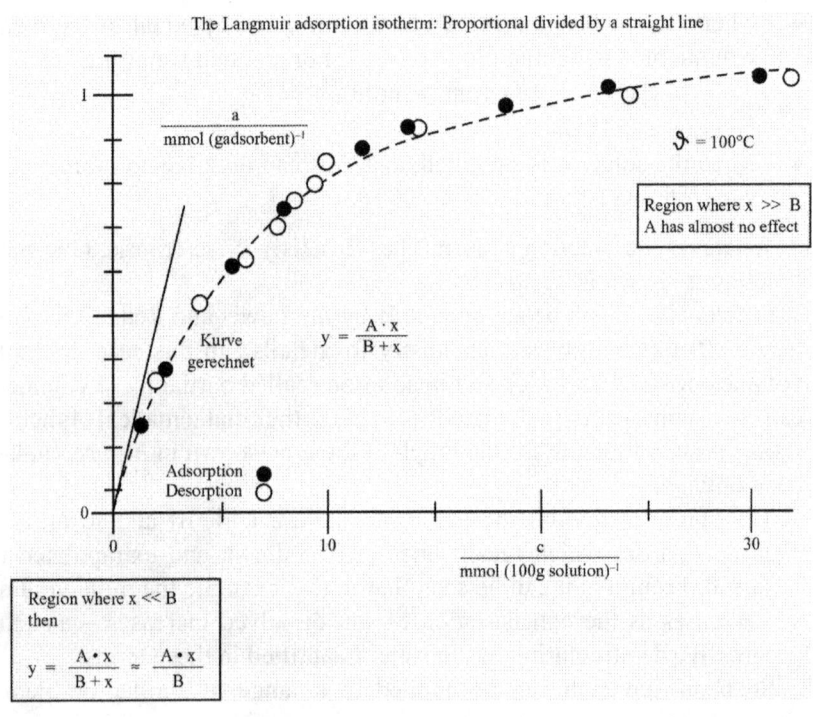

Quadratic Functions

a mmol (g Adsorbens)$^{-1}$	c mmol (100 g solution)$^{-1}$
Adsorption	
0.25	1.4
0.38	2.6
0.61	5.5
0.74	8.0
0.88	11.8
0.93	13.9
0.98	18.5
1.02	23.2
1.05	30.6
Desorption	
1.05	32.0
1.00	24.2
0.92	14.4
0.85	10.0
0.80	9.5
0.76	8.4
0.70	7.8
0.63	6.1
0.53	4.0
0.35	2.0

Table 3: Experimental values for absorption and desertion values.

The Application of Linear Regression

The dependence of the "surface concentration", a (which is the number of particles per surface area – of the dispersed silica in this case), of the concentration in the solution, c (which in this context usually is the number of particles per volume), has been derived by I. Langmuir.

The function, see *Figure 11*, which can be thought of as a proportional y = m • x, written here as (A • x) – the letter A often being used in adsorption studies, divided by a straight line y = m • x + c, written here as (B + x) – the letter B often being used in adsorption studies, starts at x << B with the proportional y = (A/B) • x and has a fixed end value A at x >> B. It can be linearised as follows

$$y = \frac{A \cdot x}{B + x}$$

$$\frac{1}{y} = \frac{B + x}{A \cdot x} = \frac{B}{A \cdot x} + \frac{x}{A \cdot x}$$

$$\frac{1}{y} = \frac{B}{A} \cdot \frac{1}{x} + \frac{1}{A} \qquad Eq.\ 11$$

Verify that the adsorption- and desorption measurements, *Table 3*, can be described correctly in this form.

One of the most essential aims of such adsorption experiments is the determination of the "specific surface", S_g, often measured in $m^2 \cdot g^{-1}$. The basis for this determination assumes that a surface becomes covered by a monomolecular layer that is adsorbed upon it but presupposes that the "demand of area" for each of the adsorbed molecules is known.

Quadratic Functions

The surface concentration of the molecules at total saturation in the mono-layer, a_∞, is found in the adsorption isotherm:

$$a = \frac{a_\infty \bullet c}{B + c}$$

Under these circumstances and supposing a demand of area of approximately 25 Å or $25 \bullet 10^{-20}$ m² it follows in this case
$S_g = 1.3 \bullet 10^{-3}$ mol/g and $6 \bullet 10^{23}$ molecules/mol and
$25 \bullet 10^{-20}$ m²/molecule

It results: $S_g = 195$ m²/g. S_g: specific surface area.

Show that the S_g of spherical particles, of radius r, may be determined if the surface can be explained as an outer surface; $S_g = (3/\rho)(1/r)$ (ρ: density).

The Application of Linear Regression

6.3 Exponential functions; the function of "natural growth and decay"

The operation of many processes changes with the amounts involved, and one of these amounts can be "time". Plots of the change in a value which describes in some way the process (e.g. temperature, number of organisms, total area, total volume, electrical charge etc.) against an amount of something which is thought to be relevant to the process, may produce a curve.

Such a curve can be given a function and when the derivative of this function is proportional to the value of the function itself, then the function (and the curve) is said to be exponential a characteristic of exponential functions is that their derivative is proportional to their respective function value itself. *Figure 12* shows an example of an exponential curve.

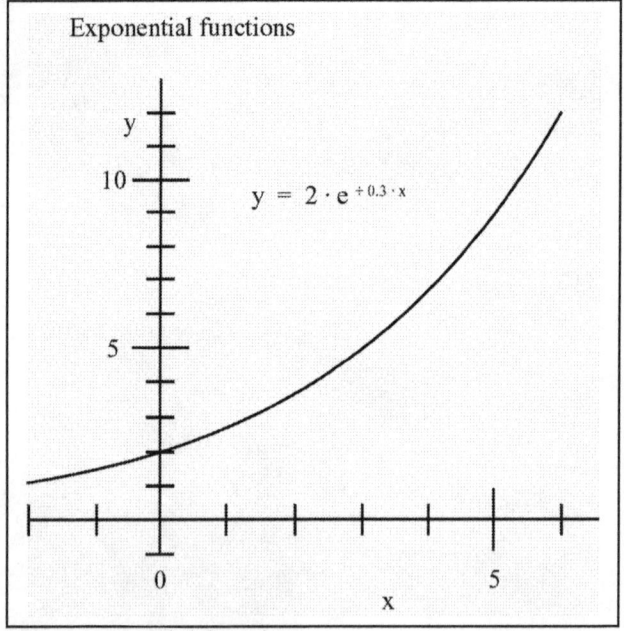

Figure 12: Exponential functions

Quadratic Functions

$y = 2 \bullet e^{+0.3 \bullet x}$ *Eq. 12* function

$\ln y = \ln 2 + 0.3 \bullet x \bullet \ln e$ **differentiation of the left side using the chain rule, firstly to y and secondly to x and recalling that ln e = 1.0**

$\dfrac{1}{y} \bullet \dfrac{dy}{dx} = 0.3$

$\dfrac{dy}{dx} = 0.3 \bullet y$ **derivate**

These curves may record either an increase or a decrease in the value of the change that accompanies a relevant amount of something to which it is linked. Positive exponentials increase and are commonly seen as descriptions of natural growth: the temporal development of biological populations has been long known to follow such trends (see the work of Thomas Maltus (1798)). Negative exponentials decrease and are commonly found to be descriptors of natural decay: the examples which follow illustrate this.

In either case, the derivative of the function has a linear relationship to the value of the function and will either increase or decrease with the value of the function. Thus, if an exponential function is suspected, check that the "growth" or "decay" rate has a linear form.

The function of growth rate is the linear form for the examination of measuring values for the presence of an exponential function at all, *see below*.

The Application of Linear Regression

Example 7: Radioactive disintegration

A: Determination of N_o, the rate or disintegration constant

To determine N_o, the rate or disintegration constant, k, and the half life, τ of a decay process, where N is the mole number and t is the time: data describing the measured values are given in *Table 4*. First, plot the data and consider what it may be telling you: it is a classical decay curve and its rate and half life should be quantifiable (*Figure 13*).

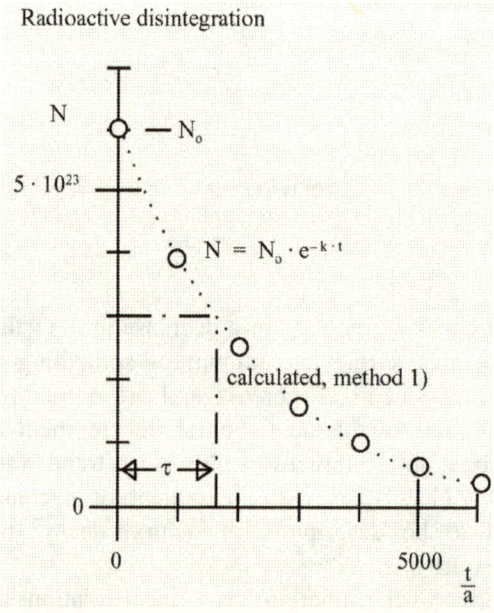

Figure 13: Radioactive disintegration

- Since the suspected function is a product made up from two factors, its logarithmic form should be a linear function between (ln N) and (t). In this case the ordinate intercept represents ln N_o, and the slope the disintegration constant with the unit for reciprocal time, in this case years, i.e. a^{-1}.

Quadratic Functions

N	$\dfrac{t}{a}$	$\dfrac{\Delta N}{\Delta t}$ a^{-1}	$\overline{N}$
N_0	0	—	
$3.883 \cdot 10^{23}$	1000		$4.953 \cdot 10^{23}$
		$-1.38 \cdot 10^{20}$	
$2.506 \cdot 10^{23}$	2000		$3.193 \cdot 10^{23}$
		$-0.89 \cdot 10^{20}$	
$1.614 \cdot 10^{23}$	3000		$2.059 \cdot 10^{23}$
		$-0.57 \cdot 10^{20}$	
$1.040 \cdot 10^{23}$	4000		$1.327 \cdot 10^{23}$
		$-0.37 \cdot 10^{20}$	
$0.671 \cdot 10^{23}$	5000		$0.856 \cdot 10^{23}$
		$-0.24 \cdot 10^{20}$	
$0.432 \cdot 10^{23}$	6000		$0.552 \cdot 10^{23}$

Using the data given in *Table 4* the result found by this method is:

$$\ln N = 54.7553 - 4.391 \cdot 10^{-4} \cdot t \qquad |r| = 0.9999999$$

Thus $N_0 = 6.025 \cdot 10^{23}$ and $k = 4.391 \cdot 10^{-4}$ a^{-1}.

- The second possible approach for evaluating the results is to make use of the proportionality between the derivative and the original function

$$\frac{dN}{dt} = -k \cdot N$$

together with the approximation (see *Table 4*)

$$\frac{\Delta N}{\Delta t} \approx \frac{dN}{dt} \qquad \text{and then} \qquad \frac{\Delta N}{\Delta t} = -k \cdot \overline{N}$$

The Application of Linear Regression

and the graphical examination on linearity. Result of the presented data:

$$\frac{\Delta N}{\Delta t} \approx +2.3424 \bullet 10^{16} - 4.321 \bullet 10^{-4} \bullet \overline{N}$$

$|r| = 0.99994$ \hfill (see *Figure 14b*)

The intercept of the ordinate may be neglected in this example because it is the constant N_o that is sought. The slope, the "rate constant", in this case is about 2% smaller than calculated by method 1. The constant N_o must be derived from the original function, for example from the figures at t = 3000a and

t = 6000a: $N_{o,\,3000\,a} = 5.9 \bullet 10^{23}$ and $N_{o,\,6000\,a} = 5.8 \bullet 10^{23}$ respectively.

Example 7: Radioactive disintegration

B: The half-life time, τ

The half-life time is that time at which the ordinate difference between t = 0 and t → ∞ has decreased to half its value. Here

$$\frac{(N_o - 0)}{2} = N_o \bullet e^{-k\tau} \qquad \text{Eq. 13}$$

It follows: $\ln 1/2 = -\ln 2 = -k \bullet \tau$ and therefore

$$\tau = \frac{\ln 2}{k}$$

Using method 1) results in $\tau_1 = 1579a$, using method 2) produces $\tau_2 = 1604a$, see *Figure 13*. The dotted curve in *Figure 13* has been calculated using method 1).

Quadratic Functions

Example 8: Concept of a cooling curve (Newton)

Not all experimental functions are as straightforward as that shown in *Example 7*. When the function describing a process is, by itself, insufficient to describe the results observed, it is necessary to consider using a more complex function. The function required to describe the cooling of water from an open container is an example of such a situation. The system looks simple but the cooling history shows that the processes involved are more complex than they appear.

The data in *Table 5* record the cooling of water with time, in a vessel of known size and shape, surrounded by air at room temperature (20.0°C to 22.5°C). Plot the data (*Figure 14a*). It is not linear although it is tempting to think of it as two or more straight line segments – but would this be reasonable?

The Application of Linear Regression

ϑ °C	$\dfrac{t}{\text{min}}$	$\dfrac{\Delta\vartheta}{\Delta t}$ °C·min^{-1}	$\overline{\vartheta}$ °C	
–	0	–	–	
82.0	1	(– 1.200)	(81.40)	(O)
80.8	2	(– 1.300)	(80.15)	(O)
79.5	3	(– 1.700)	(79.15)	(O)
78.8	4	– 0.900	78.35	●
77.9	5	– 0.820	75.85	●
73.8	10	– 0.740	71.95	●
70.1	15	– 0.620	68.55	●
67.0	20	– 0.560	65.60	●
64.2	25	– 0.420	63.15	○
62.1	30	– 0.440	61.00	○
59.9	35	(– 0.540)	(58.55)	(O)
57.2	40	– 0.370	55.35	○
53.5	50	– 0.330	51.85	○
50.2	60	– 0.238	44.85	○
39.5	105			

Table 5: Raw data for figure 14.

Quadratic Functions

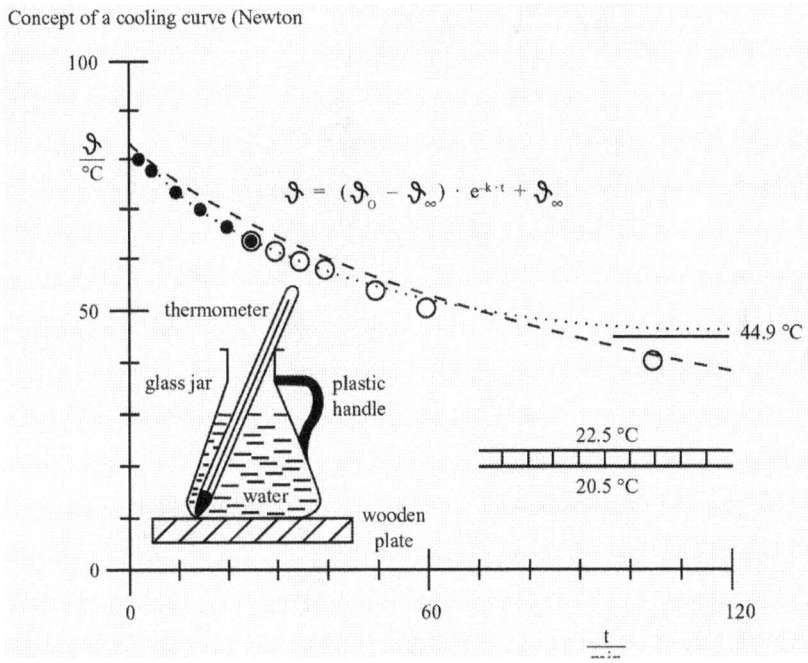

Figure 14a: Values of temperature plotted against time.

Think about the process involved: heat is leaving the system and so the loss of heat with time is the "decay" of heat, and should be expected to follow an exponential path. Further, because heat of the water is lost to the surrounding air, the temperature of the water cannot drop to values lower than that of the surrounding air. From these considerations above we can reasonably conclude that an exponential function is needed, that its derivative $d\vartheta/dt = 0$ at t_∞, and that ϑ at t_∞ = room temperature. It is therefore instructive to consider the last two facts as they have quantities attached to them, i.e. $d\vartheta/dt = 0$, ϑ_∞ at t_∞ 20.0°C to 22.5°C.

The Application of Linear Regression

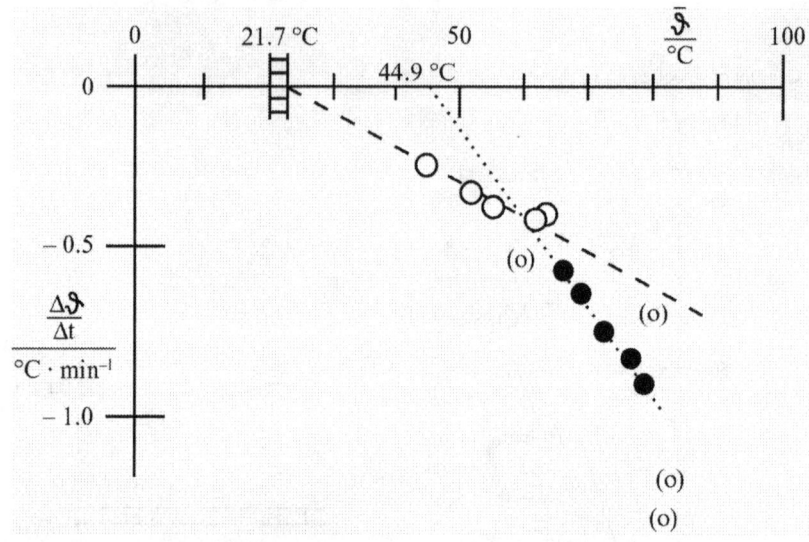

Figure 14b above is a plot of $\Delta\vartheta/\Delta t$ vs. ϑ from which at least two lines converge, i.e. there are at least two processes operating which each require an exponential function to describe them. Extrapolation of these lines produces two values for $\Delta\vartheta/\Delta t = 0$; 44.9°C is clearly wrong, and 21.7°C which is clearly possible. So, one of these processes is merging into the other, suggesting that the function required to describe the cooling overall may be obtained by adding one or more constant to the basic exponential function, i.e.

$$y = e^{b \bullet x} + c \qquad \text{Eq. 14 (compare with } Eq.\ 12)$$

Such a function cannot be linearised by taking logarithms, as was done method 1 in *Example 7* because it is not just a multiple in the exponent ($y = e^{b \bullet x}$) but contains a term of a sum (+ c). However, additive constants disappear during differentiation, as was done in method 2 of *Example 7*.

In *Figure 15* it is suggested that the approximation likely to be helpful here is:

$$\vartheta = (\vartheta_0 - \vartheta_\infty)\, e^{-k \bullet t} + \vartheta_\infty \qquad Eq.\ 15$$

Quadratic Functions

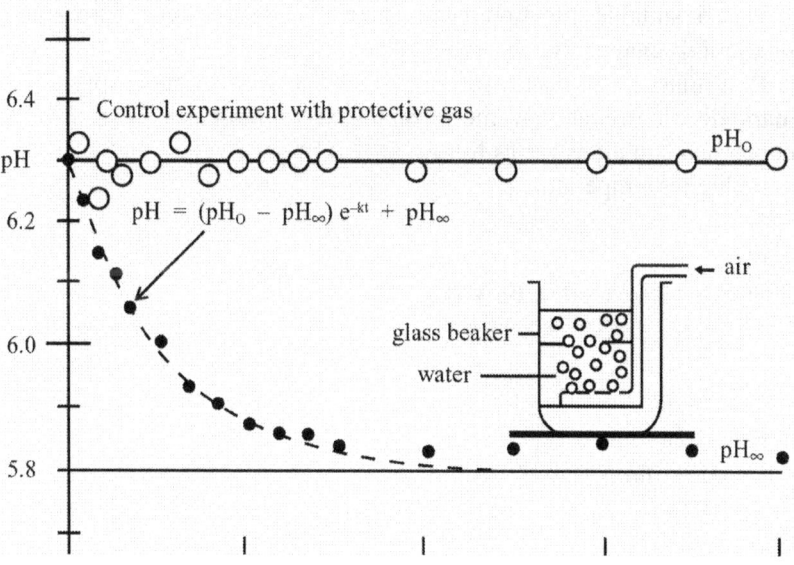

Figure 15.

This says that the temperature at any time is equal to the difference between the initial temperature and the temperature at t_∞ distributed over that time using the decay constant (k), to ϑ_∞ [this is just a curve], plus the temperature at the infinity, which is room temperature [this displaces the curve on the ϑ axis by an amount = ϑ_{room} temperature]. Differentiating *Eq. 15* yields:

$$\frac{d\vartheta}{dt} = -k \bullet (\overline{\vartheta} - \vartheta_\infty) = k \bullet \vartheta_\infty - k \bullet \vartheta$$

Note $\overline{\vartheta}$ represents the mean between two values

43

The Application of Linear Regression

Eq. 16 now contains only two constants and is open to linear regression. The third constant, ϑ_o, must be determined later, using the function eventually chosen.

Has a suitable function been chosen? Check it by comparing its predictions against the observed data.

Returning to *Figure 14b*; note that the derivations express more markedly the trends than the recalculated "direct curve" *Fig. 14a*. The concerned equations are as follows:

At higher temperatures:

$$\frac{d\vartheta}{dt} = 1.2028 - 0.0268 \bullet \vartheta$$

$|r| = 0.996$
$k = 0.0268 \text{ min}^{-1}$
$\tau = 25.9 \text{ min}$
$\vartheta_o = 82°C$
$\vartheta_\infty = 44.9°C$

and at lower temperatures:

$$\frac{d\vartheta}{dt} = 0.2323 - 0.0107 \bullet \vartheta$$

$|r| = 0.977$
$k = 0.0107 \text{ min}^{-1}$
$\tau = 64.8 \text{ min}$
$\vartheta_o = 82°C$
$\vartheta_\infty = 21.7 \, °C$

Apparently at higher temperatures heat is being lost not only by conduction across the boundaries of the system but also by evaporation from the water surface exposed at the top of the container. This combination dominates cooling at higher temperatures to around 62°C

Quadratic Functions

(see *Fig. 14b*) after which the contribution to the cooling rate from evaporation can no longer be seen.

Example 9: Dissolution of gaseous carbon dioxide in bi-distilled water at 20°C

Figure 16 illustrates an experiment where air containing CO_2 is bubbled through 400cm^3 of distilled water, at 20°C. The pH of the water is measured with time and its change reflects the dissolution of CO_2 in the water. This change is compared with the change in the pH which occurs when Nitrogen, N_2, is blown across the surface of the water, thus not allowing it to be in direct contact with air. Plot the data (*Figure 15*).

An approximation function is required for describing the change in pH with time under these conditions.

Inspection of *Figure 15* shows the limits of the function required range from pH = 6.30 to pH = 5.82 at approximately $t_{120\ mins}$, after which the pH remains constant. Thus the limits of the function required are pH_o = 6.30 and pH_∞ = 5.82.

Might the function be exponential? Here a third method for establishing the presence of an exponential and for quantifying it can be illustrated: it uses the rate constant calculated from a half-life.

The half-life is the time required for a function to pass half the distance between its two end values. Thus from *Figure 15* and *Table 6* the first half life occurs when pH ≈ 6.05; i.e. approximately twenty minutes.

If pH = 6.05 is now taken as the new upper limit and pH = 5.82 remains the lower limit, then the half-life occurs when pH ≈ 5.94, i.e. approximately twenty minutes later; the same as the first half-life.

If this calculation is repeated to obtain the third half-life, twenty minutes is again found, and so on.

Thus it can be concluded quickly and simply that the results describe a process that can be quantified by an exponential. Like *Example 8*, the value for y_∞ = the lowest values obtained, thus in this case the function will take the form of:

$$pH = (pH_o - pH_\infty) \bullet e^{-k \bullet t} + pH_\infty \qquad (Eq.\ 16)$$

The Application of Linear Regression

The rate constant (k) is easily calculated from the half life time (τ) mentioned above because $\tau = \ln 2/k$ (see *Eq. 13*) thus twenty minutes equals 0.69315 divided by k. It is $k = 0.0347$ min^{-1}. The complete original function is therefore given as follows:

$$pH = (6.3 - 5.8) \bullet e^{-0.0347 \bullet t} + 5.8$$

Example 10: Temperature dependence for rate constants, vapor pressures and equilibrium constants

Many experiments, and many processes, reveal the presence of an inflection point. Such an event can have many causes. *Fig. 16* illustrates a general form for such data and a function that may be suitable for data of this sort.

$$y = A \bullet e^{-\frac{B}{x}} \qquad (Eq.\ 17)$$

The function type in question might be exemplified for the vapour pressure over liquids or solids.

$$p_{H_2O} = p_{H_2O,\infty} \bullet e^{-\frac{\Delta H_V}{R \bullet T}}$$

ΔH_V represents the evaporation heat, R the general gas constant, numerical value 1.982 cal $\bullet$ mol^{-1} $\bullet$ K^{-1} and T the absolute or thermodynamic temperature.

Quadratic Functions

Determine applying the Gauss method, from the experimental figures, see table inserted in *Figure 16* the values of the constants, especially that of ΔH_V. Principally the method to linearise this function is the same as in *Example 7*. The difference being that the independent variable in the exponent is not just x, in this case T, but 1/T.

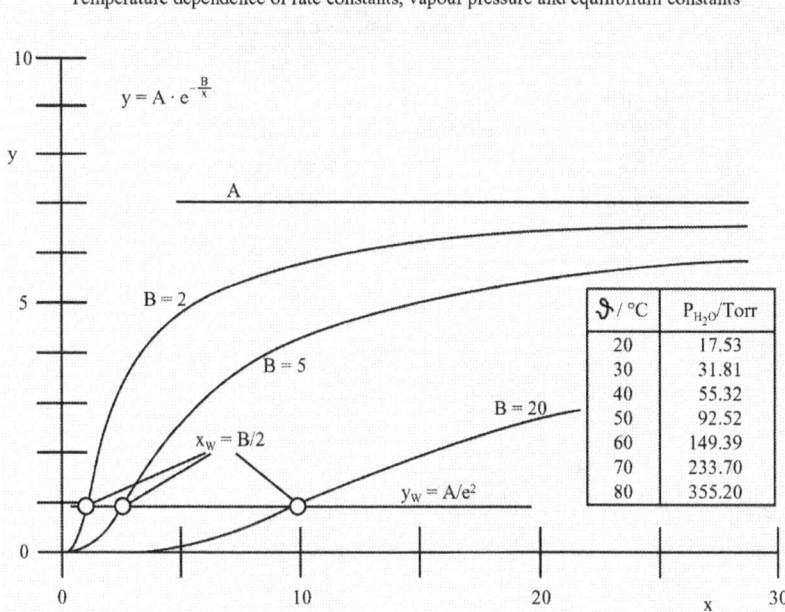

Temperature dependence of rate constants, vapour pressure and equilibrium constants

To linearise it is thus necessary to plot p_{H_2O} vs. 1/T.

By doing this $p_{H_2O, \infty}$ is found as the intercept of the plot and

ΔH_V (the evaporation heat) can be won from the value for the slope of the straight line by knowing R and how these magnitudes are linked together.

The concerned diagram ln p_{H_2O} versus 1/T gives the essential overview and allows to find the constants described above.

The Application of Linear Regression

After having determined these numerical values compare the vapour pressure, won by calculation, at 100°C and the experimentally found pressure. Where does the discrepancy come from? Determine also the co-ordinates of the inflection point under the circumstances given in this case.

To find the inflection point the following needs to be done:
- The original exponential function needs to brought into a form that allows it to be differentiated. This is done by linearising it with the help of logarithms.
- After this the linearised function has to be differentiated twice and the second derivative has to be set to equal zero. The result of this is given in Figure 16 und is $y_{\text{inflection point}} = A/e^2$.

In detail:

$$y = A \bullet e^{-\frac{B}{x}}$$

$$\ln y = \ln A - \frac{B}{x}$$

$$\frac{1}{y}\frac{dy}{dx} = 0 + \frac{B}{x^2}$$

$$\frac{dy}{dx} = + \frac{B}{x^2} \bullet y \qquad\qquad (Eq.\ 18)$$

Quadratic Functions

$$\frac{d^2y}{dx^2} = -2 \cdot \frac{B}{x^3} \cdot y + \frac{B}{x^2} \cdot \frac{dy}{dx} \quad (Eq.19)$$

Introducing equation 18 in Eq. 19 it follows

$$\frac{d^2y}{dx^2} = [-2 \cdot \frac{B}{x^3} \cdot y + \frac{B^2}{x^4}] \cdot y$$

Rewritten in a different form

$$\frac{d^2y}{dx^2} = [-2 + \frac{B}{x}] \cdot \frac{B^2}{x^4} \cdot y$$

Now this second derivation is set to zero, i.e.

$$\frac{d^2y}{dx^2} = 0$$

It thus follows for the point of inflection

$$[-2 + \frac{B}{x_{\text{inflection point}}}] = 0$$

$$x_{\text{inflection point}} = \frac{B}{2}$$

The Application of Linear Regression

The y value for the inflection point is thus

$$Y_{\text{inflection point}} = A \bullet e^{-\frac{B \bullet 2}{B}} = \frac{A}{e^2}$$

Tools to make your work easier, from QED Software

FX ChemStr – a new tool that lets you draw Chemical Structures – Without Drawing! AND…inside Word

Chemical structures are extremely difficult to draw using standard drawing tools and even specialised tools can be slow and cumbersome. What teachers and authors need is a quick way of drawing chemical structures - and now they've got it with FX ChemStruct. All sorts of chemical structures and equations are possible. Just type the structure you want, select the display option you require and a few clicks later you will have publication quality diagrams for your document. **No "drawing" is required.**

For example, If you want to draw a simple structure like acetic acid

THIS IMAGE TOOK 12 SECONDS TO CREATE!

FX ChemStruct makes this easy. Simply push the FX ChemStruct button on your toolbar and type **ch3cooh** and push the green tick to return to Word. And THAT'S ALL! You can show bonds, electrons and isomers just as you want – typing text and a few simple clicks to chose your display option.

FX ChemStruct works with Microsoft Windows packages and most effectively with Word ^{TM}, installing itself as a toolbar option so that you simply click a button to switch between writing questions or your masterwork and drawing the accompanying proof or illustration. Easy to install and easy to use! Don't believe us - try it

FREE 30 DAY TRIAL: DOWNLOAD NOW FROM WWW.TARQUINBOOKS.COM

Fast Equations for Chemistry & Physics

You type text…and you get equations… integrated with Word, Powerpoint or Excel.

FX Chem 2

If you need this equation in your document

$$MnO_4^-(aq) + 8H^+(aq) + 5e^- \rightleftharpoons Mn^{2+}(aq) + 4H_2O_{(l)}$$

you will have to spend a large amount of time selecting, formatting, sub-scripting, super-scripting and italicising; not to mention the arrows! FX Chem takes all the effort out of the job. Simply load FX Chem, type
`mno4-(aq) + 8h+(aq) + 5e- < > mn2+(aq) + 4h2o(l)`
and press enter. No formatting is required! You do not even have to capitalise the elements! FX Chem takes the information you have entered and formats it into the equation above. The only thing you have to do is to remember that => produces a single arrow and <> a double arrow.

FXPhysEquate

Imagine you want to include $\alpha = \dfrac{e^2}{2hc\,\varepsilon_0}$ in your document. To produce this equation in most equation editors is a tedious process involving menus and mouse clicks. In FX PhysEquate, all you need to do is push the FX PhysEquate button installed *within* Microsoft Word then
- Type al=e2/(2hc ep\0) watching FX PhysEquate format your equation as you go
- Click outside the equation

If you want to edit the equation later on all you need to do is click on it.

FX SciSpell...
a **NEW** science spellchecker for Microsoft Word that adds more than 17000 words and proper names to your Word dictionary making accurate work as easy as *right click and choose*. With licence prices as low as £3 for an individual, £25 for a site and £50 for site and student use (all plus VAT), it's exceptional value. And look out overleaf for our bundle prices.

Full details of all these products see Science on www.tarquinbooks.com

www.ingramcontent.com/pod-product-compliance
Lightning Source LLC
Chambersburg PA
CBHW031423040426
42444CB00005B/689